SHUTTLE, HOUSTON